AF228323

Exploring Robotics

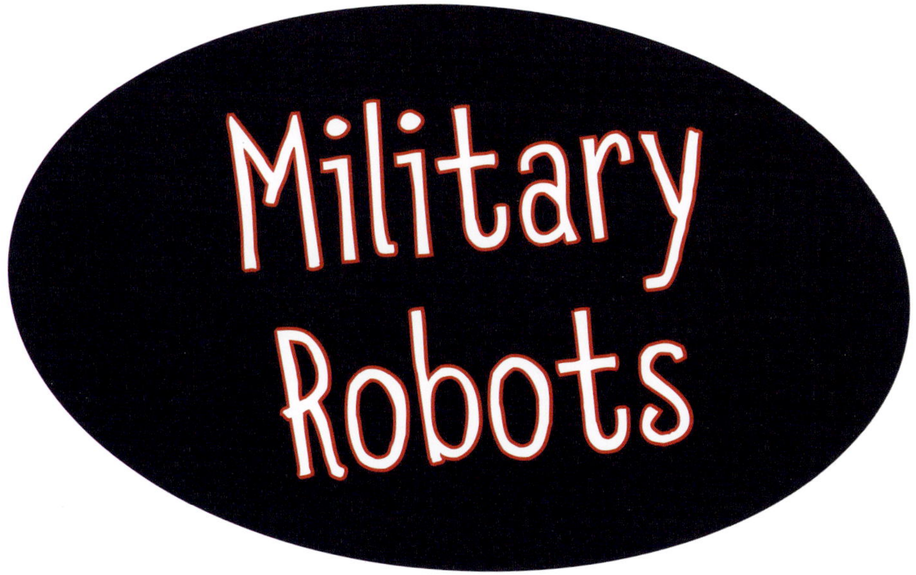

Lisa Idzikowski

Lerner Publications ◆ Minneapolis

Lerner Publications Company
An imprint of Lerner Publishing Group, Inc.
241 First Avenue North
Minneapolis, MN 55401 USA

For reading levels and more information, look up this title
at www.lernerbooks.com.

Main body text set in Adrianna Regular.
Typeface provided by Chank.

Editor: Brianna Kaiser

Library of Congress Cataloging-in-Publication Data

Names: Idzikowski, Lisa, author.
Title: Military robots / Lisa Idzikowski.
Description: Minneapolis : Lerner Publications, [2024] | Series: Searchlight books— exploring robotics | Includes bibliographical references and index. | Audience: Ages 8–11 | Audience: Grades 4–6 | Summary: "Military robots are used in the air, on land, and under the sea. Discover the history of military robots, the kinds of robots used in different military branches, and the future of military robots"— Provided by publisher.
Identifiers: LCCN 2022035736 (print) | LCCN 2022035737 (ebook) | ISBN 9781728476780 (library binding) | ISBN 9798765600078 (ebook)
Subjects: LCSH: Military robots—Juvenile literature. | United States—Armed Forces— Robots—Juvenile literature.
Classification: LCC UG450 .I39 2023 (print) | LCC UG450 (ebook) | DDC 623—dc23/ eng/20220726

LC record available at https://lccn.loc.gov/2022035736
LC ebook record available at https://lccn.loc.gov/2022035737

Manufactured in the United States of America
1-52259-50699-9/27/2022

Table of Contents

ROBOTS IN THE MILITARY

A small boxy object with four wheels hits the ground.
This machine is about 8 inches (20 cm) long and 7
inches (18 cm) wide. It can slip under cars and drive over
rocks and dirt. Soldiers can even throw it down a flight
of stairs. But it won't get hurt as a person would. It's the
SIGYN Mk1 Recon System, a military robot. Soldiers can
toss the throwable robot almost anywhere to help them
look for danger. Cameras on the robot look all around
and send pictures back to the soldiers.

Military robots, like all robots, are machines. Robots have sensors. Cameras, microphones, and other equipment help robots see, hear, and feel. After sensing the world, a robot acts. The machine's programs use information from its surroundings to tell the robot what to do. Then actuators, such as motors and wheels, make it possible for robots to move.

Robots in Control

Military robots can be big or small, short or tall. Different shapes and sizes allow military robots to accomplish all types of important jobs. Some robots take pictures and gather information. Some search for people or rescue

DURING TRAINING, AN M58 WOLF RELEASES SMOKE TO COVER THE TRACKS OF MILITARY MEMBERS.

the injured. Others deliver supplies and gear. And some robots can keep watch or hunt for unsafe objects.

Roboticists create and program robots. They fit robots with a brain. This computer tells the machine what to do. Some robots can act on their own. Others are controlled by people.

Airman First Class Stephen Leblanc controls a robot that could be used to remove an explosive.

The Fly Eye is a mini flying military robot. This uncrewed aerial vehicle, or UAV, can act on its own. Or a person on the ground can guide it. Two onboard cameras accomplish its mission—to send information by video back to the soldiers.

Key Figure

Robots do many things that help society, but they are still just machines. Machines need people to keep them running smoothly. A robotics technician is a key figure when it comes to keeping robots running. These experts perform different tasks, including designing, testing, and fixing robots. If a military robot such as a UAV isn't working, a robotics technician must find the problem and solve it.

Robotics technicians in the military, such as Dana Yowchuang (*left*), make sure military UAVs work correctly.

EARLY MILITARY ROBOTS

In 1898, an odd-looking boat motored across a small indoor pond. Crowds stared in disbelief, wondering how the boat could move all on its own. Inventor Nikola Tesla had created a new way of controlling machines. He used radio waves to steer his boat. Radio waves are electromagnetic waves used for communication. Tesla's boat was one of the first remote-controlled machines.

Wars Launch Early Robots

During World War I (1914–1918), the US military worked
on many projects, including uncrewed aircraft. The
Kettering Bug was one of them. It carried a large bomb
that weighed about 180 pounds (82 kg). The Bug could
be set so that the engine turned off during flight. When
the engine stopped, the plane would fall from the sky.
As the plane hit the ground, the bomb would explode. A
US airplane company built about fifty Bugs. But the war
ended before any of these aircraft saw battle.

Militaries experimented with other remote-controlled machines. During World War II (1939–1945), German engineers created the Goliath. The Goliath looked like a small tank. No people were on board. Soldiers operated it from a distance. They used a joystick that moved cables to steer the machine. Slow and steady, it moved close to an enemy tank or camp. Then it exploded.

A soldier inspecting Goliath

The Soviet Union, which included Russia and fourteen other countries, also produced a remote-controlled machine during World War II. The Teletank was built like a full-sized tank. It carried several weapons. Soldiers guided the machine by radio from over a half mile (0.8 km) away. Being far from battle kept the soldiers safe.

Technology Advances

Robotics technology improved over time. So did military gear. But troops in the field needed a better way to secure information. Spy planes can be shot down. Their pilots and crew can be hurt or killed. The military thought drones may be the answer. Soldiers already shot at drones for target practice.

Pilots fly fighter planes during World War II. Uncrewed drones have sometimes been used instead of planes with pilots during wartime.

The Firebee first flew in 1962. Engineers made the drone into a useful recon aircraft. They added cameras and other tools. The new craft became known as the Lightning Bug. Remote-control signals and computer programs guided these machines. They flew missions during the Vietnam War (1955–1975).

Many think that military robots began with Nikola Tesla's boat. But the march toward modern military robots took hold because of the success of early models.

Nikola Tesla, shown here in the late 1800s, created one of the first remote-controlled machines.

BY AIR, BY LAND, AND BY SEA

All branches of the US military use military robots. They help people with dull, dirty, or dangerous tasks. Some machines study the skies. Others search or stand guard on land, while others explore underwater.

In the Air

When the US military invaded Iraq in 2003, many soldiers fought on the ground. Only a few drones were used, and there were no ground robots. Five years later,

the number of ground robots in service topped about twelve thousand. The number of military drones in service keeps increasing.

Drones of all sizes fly for the military. The Black Hornet PRS is the smallest military drone in the world. Soldiers find it easy to use. It's only 6.6 inches (16.8 cm) long and weighs under 33 ounces (936 g). Almost silent, the Black Hornet PRS is a tool for sneaking around or studying soldiers on the battlefield.

The Global Hawk is a large drone. People say it looks like a beluga whale wearing a jet pack. It flies high—over 60,000 feet (18,288 m). And it can cruise for over thirty hours without landing. The autonomous UAV gathers intelligence and provides surveillance of large areas. This data gives military commanders needed information. After pilots program the Global Hawk, it takes off, flies, and lands on its own.

A GLOBAL HAWK LANDS AT THE GRAND FORKS, NORTH DAKOTA, AIR FORCE BASE IN 2021.

STEM Spotlight

The US military began using robots during combat for the first time in 2002 in Afghanistan. People believed that robots such as Hermes could help keep US forces out of danger. Hermes had cameras attached to it. The cameras surveyed caves and scouted for land mines and booby traps. Then it sent images back to troops from the Eighty-Second Airborne Division.

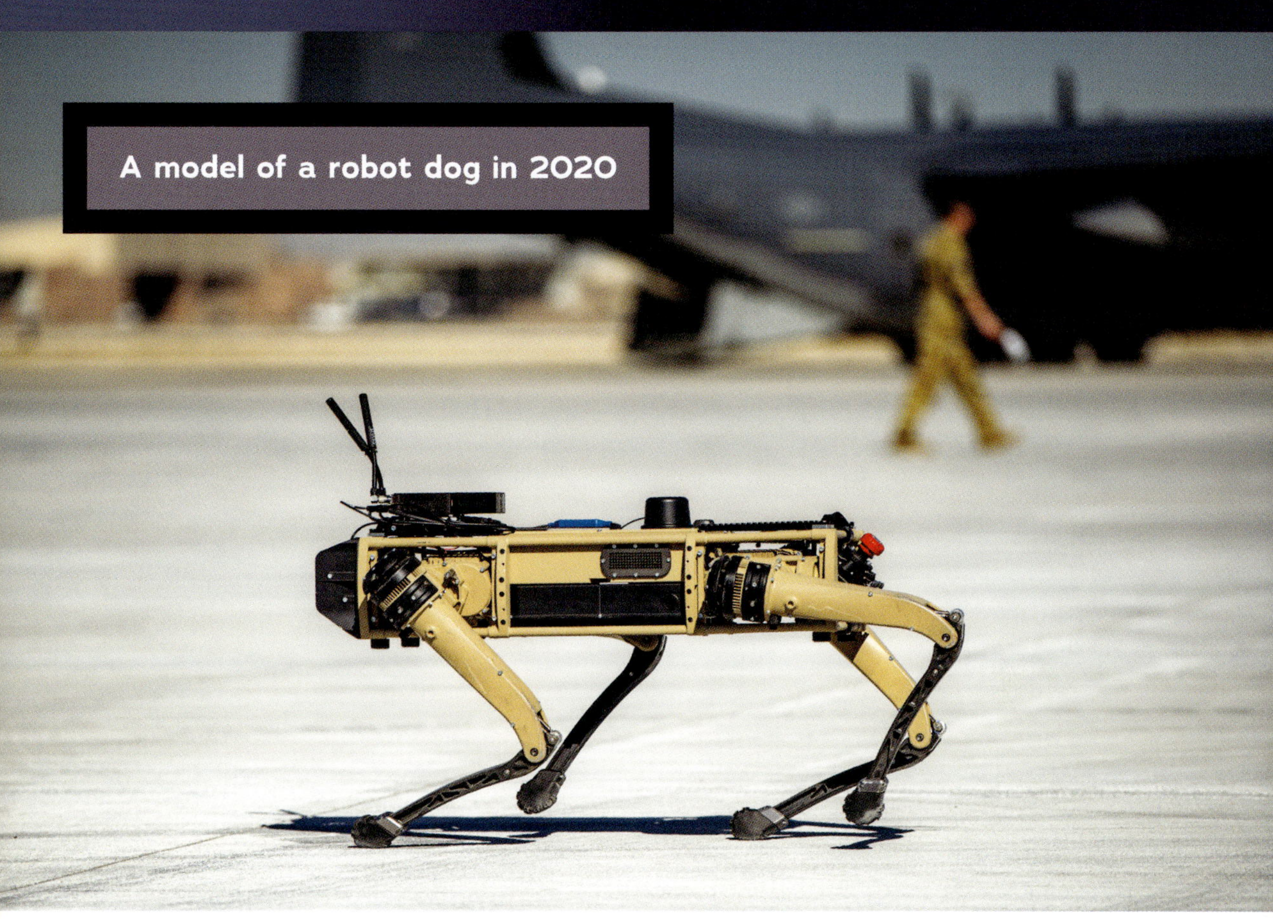

On Land

In 2020, the US Air Force used a group of semiautonomous robots in training exercises. These four-legged machines are known as robot dogs. They won't replace working military dogs that are trained to track explosives and do search and rescues. Instead, the robot dogs will be an extra set of eyes and ears on patrol on land. These machine pups will work in hard-to-reach areas. Swamps, freezing cold, and high heat are no match for robot pups.

In the Sea

Uncrewed underwater vehicles, or UUVs, gather intelligence and perform inspections. These robots can be autonomous or remote controlled. The Swordfish is a small torpedo-shaped UUV. Soldiers launch it from a boat. The sensors of the Swordfish sweep the watery environment. Soldiers use the data it collects to explore unknown areas or make maps. It can also spot mines and other dangerous objects.

Chapter 4

MARCHING ON

People in the military and people who work in robotics are trying to improve machines. They want new robots that are faster, do more jobs, and protect soldiers in dangerous situations.

The fast, quiet Orca is a robotic submarine of the future. It will follow programmed directions. But it will also act completely on its own. The Sea Hunter is a warship. In years to come, it will sail the seas with no people on board and hunt for enemy submarines.

What do some birds, fish, and insects have in common? They live and work together in groups, or swarms. Roboticists have been creating swarms of flying robots for the military. The drones of the swarm will work together. As a group, they could accomplish more than one large drone acting alone. And if one drone or a few drones in the swarm stop working, others in the group will take over.

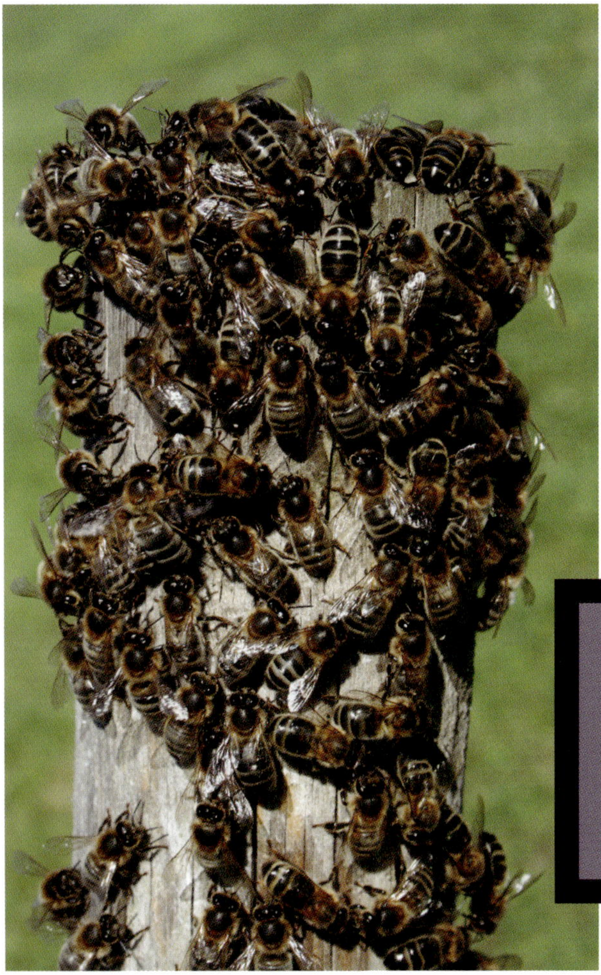

Roboticists are inspired by the way swarms of animals and insects, such as these bees, work together.

A Big Question

Some people worry that robots will become too smart. Others don't think that matters. Roboticists are building machines that use artificial intelligence, or AI. Machines with AI are getting smarter. Some scientists think these machines could even become as smart as people, or even smarter.

New and improved military robots could give an army an advantage in battle. Better machines could keep soldiers safer. Smarter machines could find enemies without getting directions from people. But advanced robots with AI could also be dangerous. Most military robots are not fully autonomous. They cannot decide on their own to take deadly action. Humans still make this decision. But what if humans weren't making that decision?

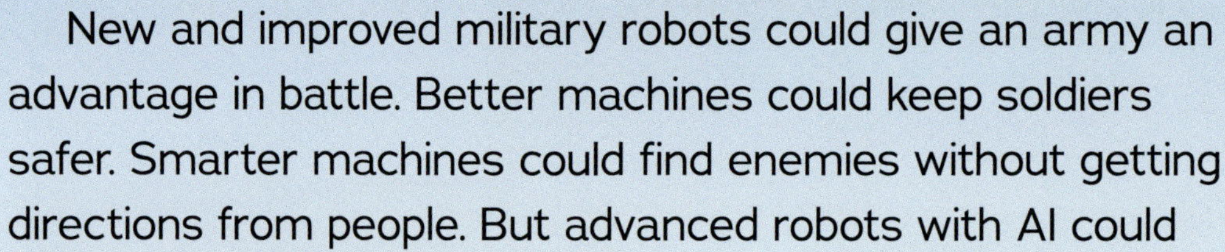

Fully autonomous weapons are known as killer robots. These robots could decide to take deadly action. Many people worry about these machines. Over a thousand scientists and experts agree. They signed a letter asking countries to ban killer robots. Many governments and militaries are struggling to balance the advantages of advanced robot technology with the dangers and problems it could cause.

Key Figure

British physicist Stephen Hawking, shown here in 2008, studied the universe. He made great achievements in science. His work on black holes brought him fame and honors. Before his death in 2018, Hawking expressed his concern about killer robots and signed the letter with other scientists and experts to have them banned.

Like all technology, military robots will continue to improve. Military leaders want robots that can help service members with all sorts of jobs. They also want robots that will help keep soldiers and civilians safe.

Glossary

actuator: a tool for moving or controlling something

artificial intelligence: the ability of a computer or robot to imitate intelligent human behavior

autonomous: able to act on its own

civilian: a person who is not serving in the armed forces

drone: a flying robot

intelligence: the ability to learn and understand

joystick: a control for a device

recon: a survey of enemy territory to get information

roboticist: a person who designs and builds robots

surveillance: a close watch of someone or something

Learn More

Britannica Kids: Artificial Intelligence
   https://kids.britannica.com/kids/article/artificial-intelligence/390648

Doeden, Matt. *Cutting-Edge Military Tech*. Minneapolis: Lerner
   Publications, 2020.

National Geographic: What Is a Robot?
   https://education.nationalgeographic.org/resource/what-robot

Rathburn, Betsy. *Drones*. Minneapolis: Bellwether Media, 2021.

Robots
   https://www.military.com/topics/robots

Silva, Sadie. *Robots in Defense*. New York: Cavendish Square, 2022.

Index

Photo Acknowledgments

Image credits: Spc. John Russell/DVIDS, p. 5; Spc. Hubert Delany/DVIDS, p. 6; Tech. Sgt. Lionel Castellano/DVIDS, p. 7; Timothy Sandland/DVIDS, p. 8; Sgt. Donna Hickman/DVIDS, p. 9; Niday Picture Library/Alamy Stock Photo, p. 11; Sueddeutsche Zeitung Photo/Alamy Stock Photo, p. 12; pablofdezr/Shutterstock, p. 13; Print Collector/Contributor/Getty, p. 14; NASA/DIVDS, p. 15; Roger Viollet/Contributor/Getty, p. 16; Cpl. Thor Larson/DVIDS, p. 18; Senior Airman Dakota LeGrand/DVIDS, p. 19; Airman 1st Class Dwane Young/DVIDS, p. 21; Petty Officer 1st Class Peter Lewis/DVIDS, p. 22; Sabena Jane Blackbird/Alamy Stock Photo, p. 24; Luke Allen/DVIDS, p. 25; Petty Officer 2nd Class Thomas Gooley/DVIDS, p. 26; Wolfgang Kumm/dpa/Alamy Live News, p. 27; NG Images/Alamy Stock Photo, p. 28; John Williams/DVIDS, p. 29.

Cover credit: Jason Wilkinson/DVIDS.